Review of the National Defense Intelligence College's Master's Degree in Science and Technology Intelligence

Committee for the Review of the Master's Degree Program for Science and Technology Intelligence Professionals

Division on Engineering and Physical Sciences

NATIONAL RESEARCH COUNCIL
OF THE NATIONAL ACADEMIES

THE NATIONAL ACADEMIES PRESS
Washington, D.C.
www.nap.edu

THE NATIONAL ACADEMIES PRESS **500 Fifth Street, N.W. Washington, DC 20001**

NOTICE: The project that is the subject of this report was approved by the Governing Board of the National Research Council, whose members are drawn from the councils of the National Academy of Sciences, the National Academy of Engineering, and the Institute of Medicine. The members of the committee responsible for the report were chosen for their special competences and with regard for appropriate balance.

This is a report of work supported by Contract No. HHM402-10-D-0036DO#2 between the Defense Intelligence Agency and the National Academy of Sciences. Any opinions, findings, conclusions, or recommendations expressed in this publication are those of the author(s) and do not necessarily reflect the view of the organizations or agencies that provided support for the project.

International Standard Book Number **978-0-309-22054-5**
International Standard Book Number **0-309-22054-8**

Limited copies are available from

Division on Engineering and Physical Sciences
National Research Council
500 Fifth Street, N.W.
Washington, DC 20001
(202) 334-3111

Additional copies are available from

The National Academies Press
500 Fifth Street, N.W.
Lockbox 285
Washington, DC 20001
(800) 624-6242 or (202) 334-3313
(in the Washington metropolitan area)
http://www.nap.edu

Printed in the United States of America

THE NATIONAL ACADEMIES

Advisers to the Nation on Science, Engineering, and Medicine

The **National Academy of Sciences** is a private, nonprofit, self-perpetuating society of distinguished scholars engaged in scientific and engineering research, dedicated to the furtherance of science and technology and to their use for the general welfare. Upon the authority of the charter granted to it by the Congress in 1863, the Academy has a mandate that requires it to advise the federal government on scientific and technical matters. Dr. Ralph J. Cicerone is president of the National Academy of Sciences.

The **National Academy of Engineering** was established in 1964, under the charter of the National Academy of Sciences, as a parallel organization of outstanding engineers. It is autonomous in its administration and in the selection of its members, sharing with the National Academy of Sciences the responsibility for advising the federal government. The National Academy of Engineering also sponsors engineering programs aimed at meeting national needs, encourages education and research, and recognizes the superior achievements of engineers. Dr. Charles M. Vest is president of the National Academy of Engineering.

The **Institute of Medicine** was established in 1970 by the National Academy of Sciences to secure the services of eminent members of appropriate professions in the examination of policy matters pertaining to the health of the public. The Institute acts under the responsibility given to the National Academy of Sciences by its congressional charter to be an adviser to the federal government and, upon its own initiative, to identify issues of medical care, research, and education. Dr. Harvey V. Fineberg is president of the Institute of Medicine.

The **National Research Council** was organized by the National Academy of Sciences in 1916 to associate the broad community of science and technology with the Academy's purposes of furthering knowledge and advising the federal government. Functioning in accordance with general policies determined by the Academy, the Council has become the principal operating agency of both the National Academy of Sciences and the National Academy of Engineering in providing services to the government, the public, and the scientific and engineering communities. The Council is administered jointly by both Academies and the Institute of Medicine. Dr. Ralph J. Cicerone and Dr. Charles M. Vest are chair and vice chair, respectively, of the National Research Council.

www.national-academies.org

COMMITTEE FOR THE REVIEW OF THE MASTER'S DEGREE PROGRAM FOR SCIENCE AND TECHNOLOGY INTELLIGENCE PROFESSIONALS

DIANE GRIFFIN (NAS/IOM), *Chair*, John Hopkins School of Public Health
JULIE J.C.H. RYAN, *Vice Chair*, The George Washington University
BRIAN BALLARD, Berico Tailored Systems
WESLEY L. HARRIS (NAE), Massachusetts Institute of Technology
KENNETH KRESS, KBK Consulting
GILMAN LOUIE, Alsop Louie Partners

Staff

DANIEL E.J. TALMAGE, JR., Study Director
KAMARA E. BROWN, Research Associate
MARGUERITE SCHNEIDER, Administrative Coordinator
DIONNA ALI, Senior Program Assistant

Preface

In the spring of 2011 the National Defense Intelligence College (NDIC) asked the National Research Council (NRC) to convene a committee to review the curriculum and syllabi for its proposed master of science degree in science and technology intelligence (see Box P-1 for statement of task). The NRC was asked to review the material provided by the NDIC and offer advice and recommendations regarding the program's suitability for achieving the desired goals of the college's program. The Committee for the Review of the Master's Degree Program for Science and Technology Intelligence Professionals convened in May 2011, received extensive briefings and material from the NDIC faculty and administrators, and commenced a detailed review of the material. This report contains the findings and recommendations of the committee.

BOX P-1

Statement of Task

To provide a national-level perspective to the National Defense Intelligence College (NDIC) as it plans for a new master's degree program scheduled to start in the fall of 2011, an NRC committee will review a draft curriculum and associated materials developed by the NDIC. The committee will produce a letter report offering advice and recommendations regarding the suitability of the draft curriculum as a basis for producing desired learning outcomes for intelligence professionals who participate in the proposed Master of Science and Technology Intelligence degree program in the School of Science and Technology Intelligence.

We wish to express our appreciation to the members of the committee for their diligent and dedicated contributions to the study and to the preparation of this report. The review process was intense and required in-depth consideration of both content and process. The committee is also grateful to the Defense Intelligence Agency for its sponsorship and to the intelligence community for its active participation throughout the study. We would like to thank NRC staff members Mike Clarke, Daniel Talmage, Marguerite Schneider, Kamara Brown, and Dionna Ali for their dedication to the study and to the preparation of this report.

Diane Griffin, *Chair*
Julie J.C.H. Ryan, *Vice Chair*
Committee for the Review of the Master's Degree Program
for Science and Technology Intelligence Professionals

Acknowledgment of Reviewers

This report has been reviewed in draft form by individuals chosen for their diverse perspectives and technical expertise, in accordance with procedures approved by the National Research Council's (NRC's) Report Review Committee. The purpose of this independent review is to provide candid and critical comments that will assist the institution in making its published report as sound as possible and to ensure that the report meets institutional standards for objectivity, evidence, and responsiveness to the study charge. The review comments and draft manuscript remain confidential to protect the integrity of the deliberative process. We wish to thank the following individuals for their review of this report:

Rita Colwell (NAS), University of Maryland
Paul Gaffney (NAE), Monmouth University
Daniel Hastings, Massachusetts Institute of Technology
Robert Herman (NAE)
Darrell Long, University of California, Santa Cruz
Alton Romig (NAE), Lockheed Martin Aeronautics Company

Although the reviewers listed above have provided many constructive comments and suggestions, they were not asked to endorse the conclusions or recommendations nor did they see the final draft of the report before its release. The review of this report was overseen by Elsa Garmire (NAE). Appointed by the NRC, she was responsible for making certain that an independent examination of this report was carried out in accordance with institutional procedures and that all review comments were carefully considered. Responsibility for the final content of this report rests entirely with the authoring committee and the institution.

Contents

Acronyms and Abbreviations

AW	autonomous weapons
BSI	bachelor of science in intelligence
CIA	Central Intelligence Agency
DFD	data flow diagram
DIA	Defense Intelligence Agency
DoS	Department of State
FBI	Federal Bureau of Investigation
FDD	functional decomposition diagram
FDDC	Foreign Denial and Deception Committee
IC	intelligence community
IO	information operations
IP	information power
KSAs	knowledge, skills, and abilities
MSSI	master of science in strategic intelligence
MS&TI	master of science and technology intelligence
NGA	National Geospatial-Intelligence Agency
NIC	National Intelligence Council
NRC	National Research Council
NSA	National Security Agency
ODNI	Office of the Director of National Intelligence
S&T	science and technology
TEC	technical expertise competency

Executive Summary

The National Research Council (NRC) was asked by the National Defense Intelligence College (NDIC) to convene a committee to review the curriculum and syllabi for their proposed master of science degree in science and technology intelligence. The NRC was asked to review the material provided by the NDIC and offer advice and recommendations regarding the program's structure and goals of the Master of Science and Technology Intelligence (MS&TI) program.

The Committee for the Review of the Master's Degree Program for Science and Technology Professionals convened in May 2011, received extensive briefings and material from the NDIC faculty and administrators, and commenced a detailed review of the material. This letter report contains the findings and recommendations of the committee.

The NDIC's new degree program was found to be both needed and justified. The user population strongly supported this effort. An examination of the program revealed that it meets the four criteria referenced: it is necessary, it is unique, it conforms to general academic practice, and the administrative processes support academic freedom.

The committee's recommendations center on two general areas. First, the committee found that the biological sciences and systems engineering were underrepresented in the existing program structure. Secondly, the committee recommends that the NDIC faculty restructure the program and course learning objectives to focus more specifically on science and technology, with particular emphasis on the empirical measurement of student achievement. Given the dynamic and ever-changing nature of science and technology, the syllabi should continue to evolve as change occurs.

Detailed review comments regarding the curriculum in general and certain syllabi provided to the committee are included in the Discussion section of this report. Following that section, findings and recommendations are provided.

Chapter 1

Program, Curriculum, and Syllabi Review

BACKGROUND

The call to arms for increased educational scientific opportunities for all Americans has been sounded by many parties, including the members of the National Research Council's (NRC's) Committee on Prospering in the Global Economy of the 21st Century, characterized by Thomas Friedman as a "nonpartisan group of America's most distinguished engineers, scientists, educators and industrialists" (Friedman, 2010). These opportunities need to be extended not only to students in the normal educational pipelines but also to veterans and those in active military service. There is a clear need for increased training and education in the sciences, technology, engineering, and mathematics (STEM) at all levels in the United States. This need, so clearly elucidated by President Barack Obama in his "Educate to Innovate" campaign for excellence in STEM education in November 2009 (White House, 2009), is also faced by the U.S. intelligence community. As the world becomes more technologically advanced, the need for intelligence officers and analysts skilled in science and technology intelligence (S&TI) increases.

The committee agrees that increased S&TI capability is critical given the current and future threat environment. It offers several reasons why the United States needs analysts and intelligence officers with S&TI skills:

- The increased speed of science and technology breakthroughs;
- The globalization of science and technology (S&T);
- The convergence of various S&T disciplines (computer science, biology, physics, neuroscience, nanotechnology, chemistry);
- The impact of commercial technology and its speed of dissemination; and
- The increased capabilities of potential adversaries, including both non-state and state actors and the willingness of these parties to share with or sell to one another.

Returning veterans of the wars in Afghanistan and Iraq require additional education and training so that their experiences and skills are not lost to the nation. In light of these needs, the National Defense Intelligence College (NDIC) has created a new program in S&TI education. This effort includes the opening of a new School of Science and Technology Intelligence and the development of new educational programs (Studds et al., 2011; NDIC, 2011a).

The NRC was asked to convene a Committee for the Review of the Master's Degree Program for Science and Technology Intelligence Professionals to consider a proposed master of science degree in S&TI. The statement of task associated with this request reads as follows:

> To provide a national level perspective to the National Defense Intelligence College (NDIC) as it plans for a new master's degree program scheduled to start

> in the fall of 2011, an NRC committee will review a draft curriculum and associated materials developed by the NDIC. The committee will produce a letter report offering advice and recommendations regarding the suitability of the draft curriculum as a basis for producing desired learning outcomes for intelligence professionals who participate in the proposed Master of Science and Technology Intelligence degree program in the School of Science and Technology Intelligence.

The NDIC is an institution of higher education accredited by the Middle States Commission on Higher Education. It is authorized by the U.S. Congress to award both bachelor's and master's degrees, to conduct and disseminate research related to national security and intelligence activities, and to engage in outreach activities. The NDIC currently offers a bachelor of science in intelligence (BSI) and a master of science in strategic intelligence (MSSI). It also offers certificate programs in various regional and topical study areas. One such specialization is a certificate in foreign denial and deception (Studds et al., 2011). The NDIC serves both military and civilian members of the intelligence community. It is currently housed at the Defense Intelligence Agency (DIA) headquarters facility in Washington, D.C.

Created in 1962 as the Defense Intelligence School, the NDIC has evolved in response to the emerging needs of its constituency and as directed by its board of visitors (NDIC, 2010; NDIC, 2011a). Most recently, in 2010, it was directed to create a school of science and technology intelligence and to develop a master of science and technology intelligence degree program (NDIC, 2011a). The curriculum under review by the committee is the material developed in response to that directive.

When a new degree program is developed by the NDIC, the program must be shown to meet four distinct criteria. First, it must be shown to be necessary. In other words, the degree program must meet an unmet need that is essential for the execution of the applying agency's mission. Second, it must be shown to be unique. It cannot duplicate existing programs that could be satisfactorily used to meet the identified need. Third, it must meet the standards for graduate degree programs met by non-federal institutions. And finally, it must allow for academic freedom for the faculty and the students so that research and classroom activities are free from undue influence and are unbiased by either job or mission (Studds et al., 2011, p. 8).

From May to August 2011, the committee met and considered the material presented by the NDIC. A committee meeting was held in May, when materials were presented and the committee was able to question the faculty and administrators of the NDIC. After the meeting with the faculty and administrators, additional material was collected to provide a comparative sampling of other curricula. Because this is a unique offering in a unique setting, there is no exact replica to examine. However, curricula similar in scope and nature were reviewed, representatives of the client population were interviewed, and the program goals were examined.

Detailed review comments regarding the curriculum in general and the syllabi provided to the committee are included in the next sections of this chapter. The committee's discussion first reflects an overview of what was provided to the Committee and then the findings and recommendations follow.

DISCUSSION OF THE MS&TI DEGREE

Curriculum

A review of the curriculum reveals that it is both unique and necessary. Comments from current members of the S&TI community revealed a true dearth of structured educational programs to create leaders in S&TI for the intelligence community. There is a distinct difference in the educational needs of S&TI leaders and S&TI analysts. Whereas analysts are required to be deeply knowledgeable about one or more subject areas of science or technology, leaders are required to have a broad appreciation for S&TI as a whole, to understand how S&TI contributes to national security, and to know how to be effective managers within the structure of the U.S. government. It was noted that the craft of S&TI is very different from simply being a scientist, engineer, or technologist. The practitioner of S&TI at all levels needs to understand not only the underlying science but also how potential adversaries use science, both systematically and culturally. This fundamental distinction leads to a substantive difference between simply learning a science and learning S&TI. As such, this discussion of the proposed degree begins with the curriculum.

The mission of the master of science and technology intelligence (MS&TI) degree program falls within the greater mission of the NDIC, which is to provide educational programs to members of the U.S. intelligence community. As a program funded by the national intelligence community, NDIC's MS&TI program has both a professional and an academic duty to enrich the skill sets of intelligence professionals, with a specific focus on improving the nation's ability to collect, analyze, and provide intelligence product in the areas of foreign science and technology development. Through its core courses and concentrations, the MS&TI program sets out to address some of the challenges identified in the following statement:

> . . . our ability to identify emerging technologies and warn about the disruptive nature of foreign developments through Scientific and Technical Intelligence (S&TI) must have a broader, worldwide focus that builds upon the traditionally peer and —near-peer competitors but extends beyond to the world of non-traditional sources of technology and asymmetric adversaries. (NDIC, 2011a)

Specifically, the mission of the S&TI community is to "address threats to national security arising from globalization of science and technology; identify disruptive consequences of adversarial technology adaptations; and provide a framework for effective collection and warning (Studds et al., 2011)." It follows, therefore, that a master's degree in this area should prepare students to both contribute to the greater S&TI mission and adopt leadership roles in the community of practitioners. And in fact, the mission for this program was clearly identified as developing leaders for the field of S&TI rather than analysts.

The MS&TI program is offered for full-time in-residence students and part-time students as well. Approximately half of the students admitted to existing educational programs at the NDIC study full time. Students come mainly from the population of military officers in the O-4 and higher ranks and civilians at GS-13 or above. Given this mature body of matriculating students, the program is focused on equipping managers with a broad understanding of intelligence and giving them a S&TI-specific background rather than on just developing a cadre of nationally trained analysts (Studds et al., 2011).

In order to offer the MS&TI degree, the university population needs to have faculty and students with a strong technical background. In particular, "the student body would have characteristic prerequisites for graduate study that are significantly different from those needed for the advanced policy, political and military strategic intelligence of the University's current elective program." The students are expected to have the "technical depth, skills, and tools to understand the potential of S&T" (NDIC, 2011a).

When examined in detail, the requirements for admission to the MS&TI program are unclear. In the briefing received by the committee, it was stated that students require "undergraduate level STEM knowledge" (NDIC, 2011a) but in discussion with the faculty and administrators there was no consensus on whether the requirement for STEM undergraduate degrees or prior STEM training would be enforced. The committee feels strongly that the STEM prerequisites must be clarified, preferably in favor of requiring knowledge of STEM prior to admission.

The current educational programs at the NDIC, including the master of strategic studies in intelligence (MSSI), require that students have an active Top Secret - Sensitive Compartmented Information (TS/SCI) clearance, which enables them to access classified resources and attend classes taught at the SCI level. The requirements for clearances, the closed facility with cleared faculty, and the breadth of knowledge brought to the classroom by other students with field experience are unique aspects of the NDIC that provide a strong underpinning to the MS&TI program.

Comprising a faculty to teach a new program of study is always a challenge. Upon examination, the NDIC does plan for a faculty worthy of the program. It has been building its capabilities over several years in specialist areas, which it already provides as focus areas for students in existing degree programs. Hence, a cadre of capable and tested faculty exists. Secondly, there are plans for recruiting and expanding faculty in critical areas that will provide the capacity for competently serving the incoming student body.

The faculty appears to bring an appropriate set of knowledge capabilities to the task of offering the MS&TI program. According to the NDIC catalog (NDIC, 2011a):

> The faculty brings a wealth of knowledge and experience gained through earning advanced degrees in intelligence-related fields, and through leadership positions in the national intelligence community. They possess a wide range of expertise in Intelligence Community (IC) topics and come from varied academic and intelligence careers. Many have served on national boards and commissions to include the National Security Council, Intelligence Science Board, the Weapons of Mass Destruction Commission, and selected presidential commissions.
>
> The Defense Intelligence Agency (DIA), Department of State (DoS), Central Intelligence Agency (CIA), National Security Agency (NSA), Federal Bureau of Investigation (FBI), Foreign Denial and Deception Committee (FDDC), and National Geospatial-Intelligence Agency (NGA) are all represented on the faculty. The Air Force, Army, Coast Guard, Marine Corps, and Navy provide faculty members as service advisors for their cohorts of students enrolled in the University. CIA, DIA, NGA, NSA, and the Department of State also provide distinguished Chairs and greatly contribute to a more integrated Intelligence faculty. (NDIC, 2011a)

The requirements set forth for the faculty are that they have "specific technical backgrounds and experience [including] an in-depth understanding not only of the science, but conditions and environment of government, military, and commercial technology research, development, and deployment." Beyond this, they are also required to hold degrees suitable to the level at which they teach and advise (NDIC, 2011a). The faculty is actively engaged in curriculum development. The NDIC curriculum working group reviews all proposed course additions or changes. The process for approving new courses was described as including several distinct steps, starting with a course proposal, then conditional approval for offering the course, followed by several trial runs in special topics status, and finally, if all conditions are met, status as a permanent course offering. The NDIC requires that at least two faculty members be able and willing to teach any new course (O'Neill, 2011).

The MS&TI curriculum was developed from the intelligence community's review of competencies and needed skill sets. The Office of the Director of National Intelligence (ODNI) published a report on skills, competencies, and intelligence capabilities that served to both identify the need for the MS&TI program and develop the structure of the program (ODNI, 2008).

The program covers the global impact of S&T, the challenges of globalization, and S&T-enabled symmetry and S&T-enabled asymmetry. The School of Science and Technology Intelligence "complements the capabilities of the existing College of Strategic Intelligence by providing select science and technology focused students better opportunities for effective science-based research, focused education and interagency outreach within a science-based educational paradigm" (NDIC, 2011a, p 38).

The core curriculum provides students with a comprehensive and laudable foundation in intelligence that includes the following, as partially extracted from the NDIC catalog (NDIC, 2011a):

> *A Global Perspective* – begin with an overview of globalization and strategic intelligence issues as the context for intelligence and U.S. national security.
> *Intelligence in a Dynamic Global Environment* – focus on developing and applying forecasting and analytical skills to systematically translate world events into products that identify and fill intelligence gaps.
> *National Security* – prepare students to be conversant with strategic intelligence demands, and to understand how the executive branch coordinates intelligence policy in the context of national security planning, Congressional oversight of intelligence policy, and budgeting, and how military and diplomatic consumers contribute to and defend intelligence policies and programs in both the executive and legislative processes.
> *Integrated Skills, Competencies, and Intelligence-Related Capabilities* – Critical thinking, analytic skills, communications, and intelligence related capabilities are woven throughout the content of the curricula.
> *Engagement and Integration* – Develop Interpersonal Skills, teaches students how to Engage and Collaborate with others and how to Influence and persuade others.
> *Communication* – Oral and Written Communication, Multi-Media Communication, and Conviction.
> *Critical Thinking* – Development of Situational Awareness, use of Creative Thinking, Synthesis and Explore Alternatives.
> *Professionalism* – Apply the principles of Ethics and Integrity, Initiative, and Adaptability.
> *Mission Awareness* – Understands the Enterprise Perspective, Customer Operations and Requirements, and Policy and Directives.
> *Analysis* – Use Analytic Rigor, how to Investigate, use Collection Systems and Operations, understand the intelligence Process and Exploitation capabilities, knowledge of Analytic Tools and Methods. (NDIC, 2011a)

Before 2009, a non-thesis option was available for MSSI, but all students are now required to develop and present a master's-level thesis (Studds et al., 2011). The planned MS&TI thesis will conform to academic standards as a formal, written presentation of original research examining an S&TI-related topic in the selected specialization. Students are encouraged to pursue research that contributes to the mission of the intelligence community. Theses may be classified or unclassified (Studds et al., 2011; NDIC, 2011a)

The MS&TI program, therefore, requires the student to complete forty-three (43) total credit hours, thirty-six (36) of which are in course work and seven (7) are in thesis research. The course work is split between three categories: fifteen (15) common core courses, six (6) degree program required courses, and fifteen (15) elective/concentration courses (Studds et al., 2011; NDIC, 2011a). The five concentration areas currently include Weapons of Mass Destruction, Information Operations and Cyber Intelligence, Emerging and Disruptive Technologies, Geostrategic Resources and the Environment, and Foreign Denial and Deception.

The five courses that make up the common core are:

MCR 601. Globalization and Intelligence Issues
MCR 603. Social Analysis and the Spectrum of Conflict
MCR 607. Intelligence Reasoning and Analysis
MCR 609. The Compound Eye: Intelligence Collection
MCR 611. Intelligence and National Security Policy

These courses are all developed and available to be taught. There are faculty to teach them, established learning objectives, and structured course material.

The courses that make up the five concentration areas are listed here by concentration. In the following lists, each course is annotated according to the maturity of the course offering. The annotation A refers to a course that is available, has been taught before, and is mature in program content, although it may be modified for the MS&TI degree offering. The annotation D refers to a course that is developed but has not yet been taught and is therefore considered to still be a work in progress according to the NDIC course development processes. The annotation F refers to a course that has not yet been developed but which is intended for development at a future time.

- Weapons of Mass Destruction
  —MST 663. WMD Counter-proliferation (A)
  —MST 665. The Biological Threat (A)
  —MST 666. Space and Missile Threats (D)
  —MST 667. The Nuclear Threat (A)
  —MST 669. The Chemical and Explosive Threat (A)
  —MST 655. Advanced Conventional and Non-conventional Weapons (A)
  —MST 656. The Economics of Technology (F)

Of these courses, five, marked A, are considered to be developed, available to be taught, and mature in terms of content, learning objectives, and structured course material. One, MST 666, which is marked D, is considered to be under development, and one, MST 656 marked F, is considered to be a future development need (NDIC, 2011b).

- Information Operations and Cyber Intelligence
  —MST 680. Information Power and National Security (A)
  —MST 681. Propaganda, Persuasion, and Influence (A)
  —MST 682. Cyber Operations (A)
  —MST 683. Foreign Information and Cyber Strategies (D)
  —MST 684. Cyber Defense (A)
  —MST 687. Advanced Information Power Seminar (D)
  —MST 698H. Cyber Attack (D)
  —MST 698J. Cyber Exploitation (D)

Of these courses, four are listed as being developed, available to be taught, and mature in terms of content, learning objectives, and structured course material. The four that are listed as under development (D) are MST 683, MST 687, MST 698H, and MST 698J (NDIC, 2011b).

- Emerging and Disruptive Technologies
  —MST 653. Advanced Science and Technology Intelligence (Process) (F)
  —MST 655. Advanced Conventional and Non-conventional Weapons (A)
  —MST 656. The Economics of Technology (F)
  —MST 657. Case Studies in Technology Transfer (D)
  —MST 658. Infrastructure Vulnerability Assessment (A)

Of these courses, two are listed as being developed, available to be taught, and mature in terms of content, learning objectives, and structured course material. One of the remaining three is listed as "under development": MST 657. The remaining two, MST 653 and MST 656, are listed as future development projects (NDIC, 2011b).

- Geostrategic Resources and the Environment (Energy Resources and Power)
  —MST 672. Intelligence and the Changing Global Resource Environment (F)
  —MST 673. Geology and Intelligence (F)
  —MST 674. Nuclear and Other Alternative Energy Sources (F)
  —MST 675. Electrical Power Systems and Distribution (F)
  —MST 656. The Economics of Technology (F)
  —MST 658. Infrastructure Vulnerability Assessment (A)

Of these courses, only one, MST 658, is listed as having being developed, available to be taught, and mature in terms of content, learning objectives, and structured course material. The remaining five courses are listed as future development efforts (NDIC, 2011b).

- Foreign Denial and Deception
  —MST 660. Introduction to Denial and Deception: History, Concepts, Issues, and Implications (A)
  —MST 662. Denial and Deception: Psychological/Cultural Aspects, and National Security Decision Making (A)
  —MST 664. Denial and Deception: Adversaries, Organizations, Activities, and Countermeasures (A)
  —MST 668. Denial and Deception: Tradecraft, Tools and Methodology (A)

Of these courses, all are listed as being developed, available to be taught, and mature in terms of content, learning objectives, and structured course material (NDIC, 2011b).

Overall, as it reviewed the proposed curriculum, the committee found that the common core courses are all available and that three of the proposed specialization areas appear to be populated enough to be offered. Two of the specialization areas, Emerging & Disruptive Technologies and Geostrategic Resources & the Environment, require significant course development efforts before they will be ready to be offered to students. Findings and recommendations are found later in the chapter.

Syllabi

A detailed description of the course work of the MS&TI program was provided to the committee. The courses annotated as "on hand" or "under development" all included a course description, a description of the course's contribution to the mission, expected learning outcomes, a list of intelligence community competencies, course requirements, grading criteria, a reading list, a session outline with subjects and issues being covered, and defined learning objectives. The course content reflected in this material seems complete and thoughtfully compiled, with the exception of the structure used for the learning objectives. The committee is concerned that the learning objectives are too broadly stated and are poorly linked to the program learning outcomes.

Additionally, a close analysis of the ODNI Competency Subdirectory for S&TI, which was used as a reference material for designing the MS&TI program, reveals that much is missing from the current course content (ODNI, 2008; Studds et al., 2011, p 14). Furthermore, of the material included, the committee had significant concerns about the learning objectives. These issues are discussed here and the committee's findings and recommendations are offered in the appropriate section.

Learning Objectives

Learning objectives form the core of any educational program. Correctly and precisely stated expectations of what exactly a student should learn in a course allow not only the precise engineering of course content to achieve those goals, but they also define the prerequisite knowledge of the student, the competencies of the instructor, and the end product of a program. Each of these is worth considering separately.

Learning objectives should be stated using action words and precise outcomes, which must be empirically measurable. For example, a learning objective that is poorly formulated is this: "The student should understand how to write a paper." A much better crafting of that learning objective would be "The student will write a research paper that adheres to appropriate academic standards; the paper will include a thesis statement, background material, data presentation and analysis, and a synthesis of research findings." In the first statement of objective, there is no way to measure how well a student understands how to write a paper. In the second, not only is the expectation clearly stated but the measure of success is also clearly defined.

The problem of poorly defined learning objectives creates confusion for the program objectives and the student learning prerequisites. Consider as an exemplar, learning objective for MST 684, Cyber Defense: "Gain an understanding of the vulnerabilities of the DoD and IC cyber systems." Were this learning objective to be recrafted in active language, it could be expressed in many ways, two of which follow: (1) "Use software modeling techniques, such as functional decomposition diagrams (FDD) and data flow diagrams (DFD), to identify critical points of vulnerability in an exemplar cyber system" and (2) "Reverse engineer an identified piece of malicious software using provided tools in a laboratory environment to identify attack paths and probable target structures." Either and both of these alternatively stated learning objectives can be measurable on an objective basis. In other words, one can measure both whether a student achieved the learning objective and how well the student achieved it. Furthermore, pre- and post-education assessments are possible to measure how much was achieved. This becomes a critical quality control function: If students are uniformly not achieving the expected level of competency, it might be an indicator that prerequisite knowledge has been incorrectly set or that the teaching process is not optimized to the student population.

However, the two restatements of learning objectives differ in three important ways: first, in the prerequisite knowledge demanded of the student; second, in the capabilities they demand of the instructor; and, third, in overall learning outcomes called for by the program. These differences are critical to the administration of the MS&TI program. Both the first and second restatements are legitimate children of the original learning objective but speak to different student populations, different faculty, and different program outcomes.

The obligatory prerequisite level of competency, for learning objective 1 would need to include a basic knowledge of information system structures. For learning objective 2, however, the requisite level of competency would need to be much higher; it would include knowledge of software structures, algorithmic development methods, and attack pattern analysis. This distinction then tracks back to the admissions process: requiring specific types of prior education or training, or having an identified set of prerequisite knowledge to be gained prior to admission to the program.

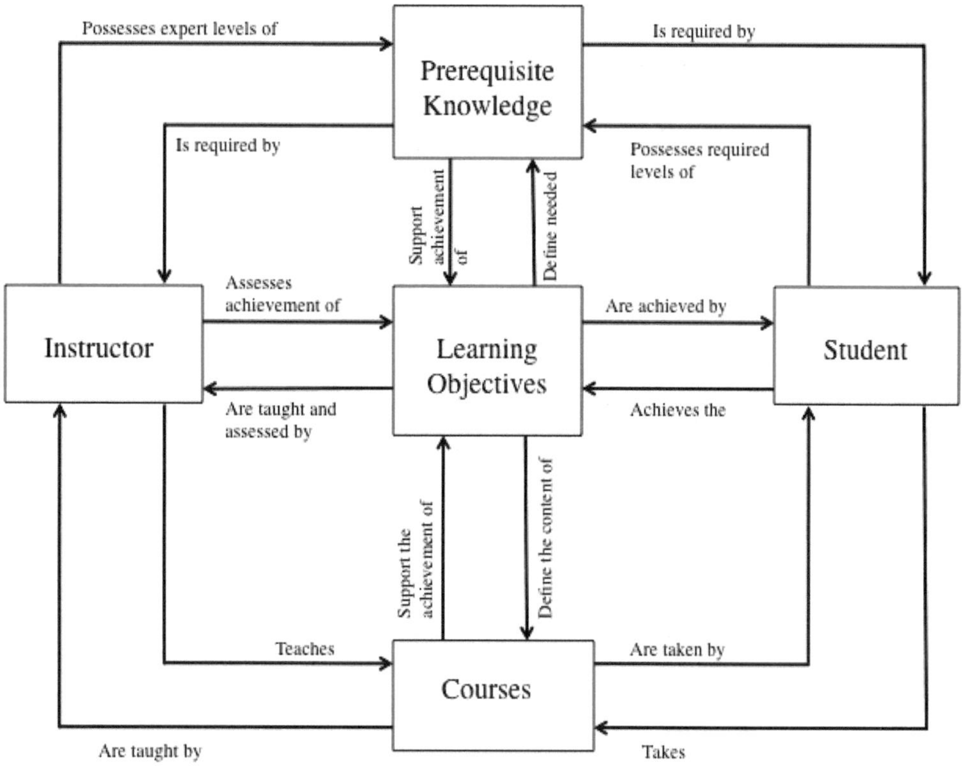

FIGURE 1-1 The importance of learning objectives.

The learning objectives influence faculty capabilities. For learning objective 1, faculty having a general knowledge of software modeling could teach the course. For learning objective 2, however, someone with significant experience in software development, reverse engineering, and cyber threat analysis would be needed to successfully teach the course. It is far easier to find an instructor to meet the needs of learning objective 1 than those of learning objective 2, which argues for a careful examination of the program learning outcomes. If these outcomes are truly to produce graduates who can reverse engineer malware, then having faculty who can teach to learning objective 1 but not to learning objective 2 will simply not do.

And it is, in fact, the overall program learning outcomes that are at stake. If the individual course learning objectives are poorly and imprecisely worded and defined, then the outcomes will range widely, depending on individual student ambitions and efforts. A more predictable graduate results from precisely defined learning objectives.

Program ODNI S&TI Competencies

A detailed breakdown of competencies and the courses in which they are covered is provided here. Two important caveats apply to the following discussion. First, it is recognized that MS&TI faculty are still developing courses, and the committee hopes that some of the comments included below will be addressed. Second, is that the committee understands that some of the syllabi have classified portions that were not shared with it. For this reason, the committee readily acknowledges that some of the following points may well be moot.

The following core competencies seem to be well addressed in the syllabi: Engagement and Collaboration, Critical Thinking, Accountability for Results, and Communication. Engagement and Collaboration includes information sharing and interpersonal skills, which are natural components of an educational environment. Critical Thinking is a priority area for the curriculum in general and as such the elements of it are addressed in each syllabus extensively. The elements of Accountability for Results that are addressed include continual learning, policy and directives, results orientation, and rigor. The elements of Communication, including information transfer and communicating with impact, are also natural elements of an academic program.

The core competencies that are not well addressed in the syllabi include Leadership and Integrity and Management Proficiency. The elements of Leadership and Integrity identified in the ODNI Competency Subdirectory, such as Developing Others and Leveraging Diversity, are difficult to teach in a classroom environment. Indeed, the syllabi emphasize integrity from an academic perspective. The elements of Management Proficiency can be taught in a classroom environment, but they compete for time with more technical subjects. When hard choices have to be made regarding use of available time in the classroom, it would be appropriate to prioritize the material at the core of a program rather than more general topics. In this case, technical knowledge, skills, and abilities (KSAs) may be rightfully prioritized over management proficiencies, depending on the students' backgrounds. That is not to downplay the importance of such proficiencies; on the contrary, they are vitally important to the desired end product of the MS&TI degree: a leader who is competent to manage an S&TI function. As discussed in findings, this shortcoming of the MS&TI program might be addressable using clever partnerships with other programs.

Sixteen (16) technical expertise competencies (TECs) are listed in the ODNI Competency Subdirectory. Those TECs that are well-covered in the MS&TI syllabi include the following: Collection Systems Capabilities, Cryptography, Customer Operations and Requirements, Intelligence Disciplines, Research and Exploration, and Researching. The TECs that seem to be adequately covered in the syllabi include the following: Processing and Exploitation Capabilities, Project Management, Technology Integration/Insertion, Tools and Methods, and Transportation. The TECs that are not well-covered, either due to incompleteness of the topic or the absence of it altogether, include the following: Contracting/Procurement, Counter-proliferation, Intelligence Topics, Systems Engineering, and Targets: General.

Given the time constraints of the program, it is probably logical to not cover everything and in this case, Contracting/Procurement is an obvious topic to leave out. However, it is appropriate to review the Counter-Proliferation TEC, described as "skill at supporting counter-proliferation operations with S&T operations" (NDIC, 2011b), for completeness. Reviewing all syllabi provided does not readily reveal any coverage of this competency. It is entirely possible that this competency is covered in MCR 601 and/or MCR 611 but, if it is, this is not explicitly stated in the syllabi and so that lack of coverage can only be conjectured. If, in fact, this competency is not covered, it may be worthwhile reviewing the syllabi to consider whether there should be more emphasis on this competency. One of the TECs that is not completely covered is Intelligence Topics. While 55 KSAs are listed under Intelligence Topics, the syllabi that were provided, which do not include sensitive KSAs, show 27 of them are not addressed. These are predominantly the KSAs that deal with biological weapons development, deployment, and testing. Because this is a critical element of the S&TI challenge today and an extremely complicated and difficult one, the NDIC should consider a more comprehensive approach to educating the MS&TI student population in biodefense capabilities. "The breadth of biological threats is much broader than commonly appreciated and will continue to expand for the foreseeable future" (Choffnes et al., 2006, p 28; NRC, 2006).

This very important area of ST&I, biological threats, appears to be under-represented in the current curriculum. This deficiency is exemplified by MST665 "The Biological Threat," which is described as "on hand" and includes two lectures on bacteria, two on viruses, and two on toxins, but the content of these lectures is not detailed. They do not appear to include basic principles of epidemiology, synthetic biology, the engineering of organisms for resistance and increased virulence, vaccines and protective immunity, or sources of new disease agents (emerging infections).

Additionally, the TEC for Systems Engineering is viewed by the committee as a critical omission. The interdependencies that characterize the global innovation landscape require for ability to understand and apply systems logic when analyzing the various bits and fragments of information collected on foreign capabilities. In particular, the integration of vastly different technologies and sciences is resulting in capabilities that can be understood only when viewed systemically. This is particularly true in the biosciences: "Other fields not traditionally viewed as biotechnologies–such as materials science, information technology, and nanotechnology–are converging with biosciences in unforeseen ways and enabling the development of previously unimaginable technological applications" (Choffnes et al., 2006, p 26; NRC, 2006; NRC, 2005). Developing these new technologies with fully open information and sharing is difficult enough; the challenge of divining what is occurring in a closed environment where clues are being gleaned from a variety of sources is vastly more difficult.

Systems engineering expertise is also useful for the analytical function from a team communications perspective.

> Useful analysis requires effective communication among diverse individuals. Members of an analytical community must communicate with each other and with their clients. Both kinds of communication can bring into contact individuals with very different missions, backgrounds, and perspectives. Within an analytical community, a single analysis might require communication among individuals with expertise in economics, anthropology, psychology, engineering, and logistics. Each contributing discipline might have subfields and competing theories, each needing to be heard. Setting the terms of an analysis and reporting its results might require communication with clients who differ from the analysts in their objectives, careers, and education. These professional differences overlay the cultural, socioeconomic, and other differences that can complicate any communication in a diverse society. (NRC, 2011a)

Expertise in systems engineering establishes an ability to "integrate and verify system elements" and analyze "processes and procedures [used] to implement operational schedules," according CIA's description for the job of systems analyst, available online at www.cia.gov/careers/opportunities/science-technology/systems-engineer.html, last accessed July 15, 2011. Furthermore, systems engineering expertise is increasingly being identified as a critical core competency that is presenting short supply in today's intelligence community (NRC, 2011a; NRC, 2011b). In sum, the Systems Engineering TEC appears to be an increasingly valuable part of the intelligence analysis equation, particularly in S&TI.

This discussion serves as the overarching framework for the findings and recommendations of the committee. It is worth restating that both the discussion and the findings and recommendations are based solely on the material presented and are made with the express knowledge that some material may not have been shared with the committee for reasons of sensitivity or classification, perhaps rendering some of the discussions or the findings and recommendations moot.

FINDINGS AND RECOMMENDATIONS

General

In general, the master's program is well thought out and gives students a solid foundation on a wide range of S&T intelligence issues. For the most part, the courses cover a lot of ground, and students will be exposed to a great deal of S&TI-related material. The course development process is inclusive and requires that at least two faculty members be available to teach a proposed course. The learning outcomes are well defined in terms of Bloom's Taxonomy. There are, however, some weaknesses, which are discussed below.

Finding 1: The Master of Science and Technology Intelligence (MS&TI) program meets the degree authorization criteria. Specifically, it is necessary, unique, meets academic standards for similar programs, and supports the academic freedom of both faculty and students.

Finding 2: The faculty appears to bring an appropriate set of knowledge capabilities to the offering of the MS&TI program.

Finding 3: While the faculty has a process in place for developing new courses, it does not appear to have a systematic way of reviewing and updating existing courses to keep the content abreast of the rapid pace of science and technology developments worldwide.

Recommendation 3: The faculty should define and institute a process for reviewing and updating the technological basis for course content to keep the material fresh as new developments occur.

Learning Objectives, Course Assessment, and Prerequisite Knowledge

In the existing syllabi, learning outcomes are described generally according to the standard terminology associated with Bloom's Taxonomy. As such, the learning is described using verbs such as "understanding" and "having an appreciation for." However, when assessing actual learning achievement, descriptions like these are not easily mapped to empirically measurable outcomes. Further, as discussed previously, the learning objectives (LO) of existing courses (NDIC, 2011b) are not uniformly stated. LOs are what students will know *and* be able to do as a result of successfully mastering the content of a course (Crawley et al., 2007). LOs must be executable by students and measurable by faculty. LOs define the prerequisite knowledge, influence the admissions processes, and ultimately describe the overall program outcomes.

Finding 4: The learning outcomes of existing courses are not uniformly stated correctly in an active achievement structure. This impacts both the expectations of student preparation prior to admission to the program, the faculty competencies, and the overall program outcomes.

Recommendation 4-1: The National Defense Intelligence College should reformulate the learning outcomes of the courses of its MS&TI program in action verbs with empirical assessment measures. Specifically, NDIC should write learning objectives that are empirically measurable, that support pre- and post-education assessment, and that contribute directly to the overall program outcome goals.

Recommendation 4-2: The reformulated learning objectives should be used to identify prerequisite knowledge requirements, which should be used as a guideline for student admission and for student remedial preparation. The committee strongly feels that the STEM prerequisite for the MS&TI degree should be clarified and should include STEM prerequisites acquired by means of undergraduate education, life experience, or self-education. The linkage of prerequisite knowledge to the learning objectives can help in this clarification process.

Recommendation 4-3: The National Defense Intelligence College should discuss the curriculum focus in consistent terms. It should clearly define the set of students that the school targets for education and the outcome expected after they complete the degree.

Recommendation 4-4: The National Defense Intelligence College should use the reformulated learning objectives to identify the required faculty competencies, which could then be used as the basis for a hiring plan and/or personnel development plans.

Closely related to this issue is the challenge of assessing the relationship between teaching and learning for each course. The process of learning requires the student to pay attention, perform assigned activities, and study; the process of teaching requires the instructor to provide material to the student, create a favorable learning environment, and assess student achievement. A strategy for assessing the teaching–learning process is critical for continuous improvement of an education program. This is most appropriately done at two levels: the course level and the program level. Assessing teaching–learning at the course level helps to identify weaknesses in preparation or unsuccessful teaching approaches and, conversely, superior programs. Assessing teaching-learning outcomes at the program level helps to identify system wide opportunities for synergy or improvement that transcend course-level concerns.

Finding 5: Assessment of the teaching-learning process is absent in course descriptions/syllabi.

Recommendation 5: The National Defense Intelligence College should define and clearly state its strategy for assessing the teaching-learning process for each course in its MS&TI program consistent with the reformulated learning objectives. It should aggregate course-level assessments into a program-level assessment so that root causes of failures or opportunities for synergy and systemic improvement are identifiable.

Finding 6: The grading criteria in some syllabi, including MCR 601, identifies the standards for letter grades such as A, A-, etc, as well as characterizes those grades in terms of expected achievement. The grade of B- is identified as a "below average grade" in some instances, which is contrary to standard academic practice.

Recommendation 6: The faculty should standardize all grading interpretations to standard academic practice, where the grades reflect the level of achievement of the learning objectives in the following gradation families: A, excellent; B, good; C, average; D, poor; and F, failure.

A significant issue that was identified in conversation with faculty members—namely, was how students come to the program in a full-time basis. Whereas military students are typically sent for education between assignments and the cost is absorbed by the parent service, for civilian employees, the model is quite different. For civilians in the intelligence community, the parent institution absorbs the cost of both the education and the absence of the employee—that is, of the unfilled position and work that is not done or must be done by other analysts for the entire time of the program.

Finding 7: Students coming to the program from the civilian ranks of the intelligence community seem to face a hurdle in getting time-off from work to attend the program in residence.

The NDIC's MS&TI curriculum addresses the important intelligence community challenges of improved S&TI leadership and more technical expertise. Our nation's S&T capabilities were once dominant but now exist in a multipolar world and face serious competition from China and Europe and many other countries such as India, Brazil, and Japan. S&T developments across the globe are priority economic and national security issues, but U.S. S&TI leadership, its intelligence analysts, and its technical depth are widely acknowledged to be inadequate. In addition, the intelligence community may devalue S&TI because of a bias for clandestine intelligence, a challenge that is exacerbated by IC S&TI leadership that does not compete well in the bureaucratic struggles for resources and influence. More trusted S&TI analysts are needed, but they are both difficult to recruit (because of commercial competition) and difficult to retain (because of the lack of respect for technical expertise within the IC). The MS&TI program, with its emphasis on S&TI leadership and technical analytic skills, is a helpful response to the growing need for better S&T leadership and more effective technical analysts.

There seems to be some discrepancy between the concept of building leaders rather than analysts and the following excerpt from the NDIC catalog: "Our curriculum focuses on developing and applying forecasting and analytical skills, sustaining and supplying scientists and engineers for national security elements, and enhancing operational capability options for our warfighters and strategic decision makers" (NDIC, 2011a). In conversations with both the faculty of NDIC and the user community, there was consensus that the focus of the program should be the development of leaders for the S&TI community (Studds et al., 2011).

Finding 8: Despite a dire need for effective S&TI leaders to grow and manage a strong S&TI community, the review of the curriculum does not find courses that are directed to the development of leaders.

Recommendation 8-1: The National Defense Intelligence College faculty should consider the development of a leadership mentoring program, using shadow assignments, rotating short internships, or special lecture series, for MS&TI students.

Recommendation 8-2: The NDIC faculty should consider exposing students to the working S&T community, including through visits to focus-relevant laboratories, to broaden their appreciation of the scope of both S&T and S&TI.

Finding 9: There is a distinct difference between the educational needs of S&TI leaders and the needs of analysts. The learning objectives and the learning outcomes specified for each course for the MS&TI program do not address the educational needs of intelligences S&TI leaders.

Recommendation 9: The National Defense Intelligence College faculty should craft program learning outcomes around the competencies of leaders and decompose these competency outcomes into learning objectives for each of the courses. For example, a competency of being able to direct group efforts could be developed through structured requirements in courses to lead group activities.

Structure

The limited time students usually expect to devote to earning a master's degree, much less than that needed for a Ph.D., is a significant constraint. Furthermore, a master's degree, despite its name, does not imply mastery of subject but rather an elevation of knowledge beyond that acquired through previous experience, such as undergraduate education, life experiences, or self-education. This reality limits curriculum options, which range from broad survey courses without a thesis requirement to reduced course work but a substantial thesis requirement. The current MS&TI curriculum calls for a thesis effort that promotes independent thinking and is useful practice for intelligence analysis and communication. In the best cases, the thesis also produces new insights and intelligence techniques. The committee supports the inclusion of a substantive thesis and accepts any consequential reduction in opportunities for course work.

Finding 10: The requirement for a thesis supports the goals of the MS&TI program, albeit at the cost of time for course material.

Recommendation 10: The National Defense Intelligence College should consider offering a capstone course that would bring all of the course material together into a simulation or "war game" environment, where the simulation is played by teams of students. The teams should be made up of students with different kinds of expertise, and the simulation should pose challenges closely associated with real world S&TI challenges, thus giving students the opportunity to internalize the challenges of operating while gaining leadership experience in S&TI matters.

The MS&TI program may need future changes. The current prerequisites involve baseline technology knowledge. In principle, this technical prerequisite eliminates the need to provide basic technical courses and keeps the program short enough to complete in one year. In the committee members' experience, however, and consistent with the flat or downward trend in the number of technical degrees awarded to U.S. citizens, the size of the pool of students that strictly possess this baseline technology knowledge is stagnant or shrinking (Engineering Trends, 2006). As a result, the number of qualified students will be difficult to maintain.

As noted in the first half of this chapter, the committee agrees that there is an increased need for intelligence officers and analysts skilled in S&TI. This increased need stems from the current dearth of qualified personnel working in intelligence, the growing technological sophistication of adversary activities, and the convergence of various S&T disciplines that are resulting in surprising innovations (ISB, 2006).

Finding 11: A consistent pool of highly qualified students is most likely going to be a problem due to the nature of the US educated populace and an overall weakness in STEM areas.

Recommendation 11: To insure a steady and even increasing supply of students, consider adding in-depth technical courses to the MS&TI curriculum.

Syllabi

A review of the syllabi provided for the core courses, the program required courses, and the courses associated with areas of concentration revealed an MS&TI program that is still very much a work in progress. Some elements of it are mature while others are in development. This may be partly because some of the courses have migrated from the existing MSSI program and partly because the program is so new. The findings and recommendations presented here refer to the existing syllabi that were presented to the committee and not to any yet-to-be-developed courses, and they do not pertain to sensitive aspects of existing courses.

The syllabi were examined with a particular eye to the ODNI S&TI competencies. Some significant S&TI competencies were found to be underrepresented, as discussed earlier.

Finding 12: The ODNI S&TI competencies are not completely addressed in the existing syllabi. One of the TECs with insufficient coverage is Intelligence Topics. This TEC is enormously broad, so it may be that the NDIC faculty plan to rectify this shortcoming in the future.

Recommendation 12: The National Defense Intelligence College should develop a matrix of competencies and derived learning objectives against which to target development of future courses.

Finding 13: The Intelligence Topics TEC was found to be incompletely covered, especially in the areas of biology and systems engineering. Indeed, some basic aspects of biology and biological threats are underrepresented in the curriculum.

Recommendation 13-1: The National Defense Intelligence College faculty should review the entirety of the syllabi, including the sensitive elements, to ensure that biological threats and S&TI needs are adequately represented in the program.

Recommendation 13-2: The National Defense Intelligence College faculty should revisit the area of Systems Engineering to ensure that all students receive adequate grounding in this TEC.

Finding 14: The critical core competencies that are not well addressed in the syllabi are Leadership and Integrity and Management Proficiency.

Recommendation 14: Given the goal of graduating leaders for the science and technology intelligence community, the National Defense Intelligence College faculty should consider the program in light of leadership needs and offer students the option to participate in a capstone simulation war game, perhaps in place of the thesis requirement.

Core Courses

There are five core courses for the MS&TI program, two of which were not available for review due to the sensitivity of their content. These five courses are MCR 601, Globalization and Intelligence Issues; MCR 603, Social Analysis and the Spectrum of Conflict; MCR 607, Intelligence Reasoning and Analysis; MCR 609, The Compound Eye: Intelligence Collection (Sensitive); and MCR 611, Intelligence and National Security Policy (Sensitive). Comments on these core courses follow.

MCR 601 is also titled "Intelligence and the Global Strategic Environment." The committee suggests that this course leverage the work performed by the National Intelligence Council (NIC) to frame the discussion on global issues. Of particular interest would be the NIC's Global Trends Forecast (available at http://www.dni.gov/nic/NIC_2025_project.html). As noted on the NIC Web site, "'Global Trends 2025: A Transformed World' is the fourth in a series of unclassified reports prepared by the NIC in recent years that takes a long-term view of the future. It offers a fresh look at how key global trends might develop over the next 15 years to influence world events." This material would be very valuable to the course. Given the velocity of global change, alternative futures, signals, and tipping points might be used to help students understand the possible futures and learn to deal with low probability, high-impact possible events.

Recommendation 15-1: The National Defense Intelligence College should incorporate the National Intelligence Council (NIC) work on global trends in appropriate courses to frame the discussion on global issues.

Recommendation 15-2: The National Defense Intelligence College should add the following course goal to MCR 601 in order to emphasize the focus on science and technology:

> Understand that science and technology and its globalization increase the potential for both threats and advancements. The credo of scientific openness that drives this globalization dispenses technical knowledge worldwide, providing opportunities for grave mischief such as the bioengineering of biological weapons using the same technology and techniques as are used for drugs and cures. In a short number of years, economic juggernauts such as Google and Facebook grew from concept to market dominance. Intelligence analysts must have the knowledge to detect and distinguish between future threats and beneficial outcomes.

Recommendation 15-3: The NDIC should add the following course goal to MCR 601:

> Understand the sources for classified science and technology innovations within the U.S. intelligence community and know where highly technical advice might be sought in each of the critical technological development areas.

MCR 603, Social Analysis and The Spectrum of Conflict, appears to be mature and well designed. Apart from the previously mentioned comments on learning objectives and an emphasis on S&TI, the committee has no further comments on this course.

MCR 607, Intelligence Reasoning and Analysis, often requires judgments based on incomplete facts. Many of us handle ambiguity with ease, perhaps as a result of survival-positive attributes such as the need to quickly identify friend or foe. Our rapid processing, however, may too often be based on beliefs that select data that reinforce our predispositions. We see what we expect and ignore the unexpected. Even with important national intelligence issues, beliefs and bias rather than the rational assignment of outcomes in the face of uncertainty may drive our conclusions. As recognized in the S&TI program, the reduction of bias in intelligence analysis is an enduring need.

Finding 16: The emphasis on reducing bias in S&T is praiseworthy. Many intelligence failures resulted from the insidious bias that comes from tribal proclivities that make us think others will do it the "American way." Within the U.S. S&TI community, we tend to have an advance technology bias and sometimes underestimate the impact of less advanced or commercially available technologies.

Recommendation 16: The National Defense Intelligence College should add the study of two specific biases that science and technology intelligence analysis may be particularly susceptible to: authority bias and abhorrence bias.

- Authority bias occurs when an analyst is more likely to believe a potential outcome is possible because of the authority of the source, even if the source is absolutely wrong or speaking outside of her area of expertise. This becomes particularly problematic when, due to rapid changes in S&T, an analyst must rely on advice from outside experts. It is very easy to believe, without questioning, outside S&T experts, especially on esoteric subjects.
- Abhorrence bias occurs when a potential outcome is discounted, either consciously or subconsciously, because it is culturally or personally abhorrent to the analyst. For example, an analyst might be inclined to reduce the probability of an abhorrent threat even if all signs are pointing to it being the most likely scenario, such as flying an airplane into a large building filled with innocent civilians from around the globe or surgically implanting explosive devices in humans to evade security.

Program Requirements

The MS&TI program requirements, aside from the common core courses and the concentration area courses, consists of MST 604, Advanced Intelligence Methods of Analysis, and MST 613, Science, Technologies and Intelligence. This section provides the committee's comments on these two courses.

MST 604, Advanced Intelligence Methods of Analysis, appears to be a mature class with well-defined objectives. Apart from the previously mentioned comments on learning objectives and an emphasis on S&TI, the committee has no further comments on this course.

MST 613 Science, Technologies, and Intelligence, would appear to be the fulcrum upon which the MS&TI pivots. As such, it is a critically important course. This is the course that is specifically focused on S&TI in the program, and its description in the syllabi clearly states that a primary purpose of the course is to teach how to identify disruptive technologies. In support of this goal, it would be worthwhile to have a deeper discussion on technology forecasting and tracking strategies in the course. Seeing into the future is complex and uncertain, yet there are techniques that have been developed to lend structure to such endeavors, particularly in the technological arena. A survey of forecasting methodologies would be worthwhile. Special emphasis should be placed on approaches to forecasting disruptive technology. Students should have an understanding of the appropriate use of forecasts and how to frame and present technology forecasts in analyses. The committee notes that a good teaching material would be the NRC's *Persistent Forecasting of Disruptive Technologies* (NRC, 2009).

Finding 17: The material included in MST 613 on forecasting is incomplete and could benefit from increased rigor, particularly with respect to disruptive technology identification.

Recommendation 17: The National Defense Intelligence College faculty should add a session on disruptive technology forecasting and tracking to MST 613. Further, the NRC report *Persistent Forecasting of Disruptive Technologies* (NRC, 2010) should be added to the materials for teaching this course.

Areas of Specialization/Concentration

There are currently five of concentration areas for the MS&TI: Weapons of Mass Destruction, Information Operations and Cyber Intelligence, Emerging and Disruptive Technologies, Geostrategic Resources and the Environment, and Foreign Denial and Deception. The courses associated with these concentration areas of are at a mix of maturity levels. Tthe courses that were presented with the syllabi and found to warrant suggestions are discussed next. Those courses that were not mature or that did warrant suggestions are not covered. The concentration Geostrategic Resources and the Environment currently has no developed courses, so no concrete commentary is possible.. However, the committee supports the inclusion of this as a concentration area in the future MS&TI program because the issues involved are so important and have the potential to impact both global and regional stability, which are important elements of national security.

Weapons of Mass Destruction

The inexorable advances in S&T increase the importance of intelligence about weapons of mass destruction. S&T discoveries give groups and individuals who scheme to advance their causes more options to harm the United States. As an example, the knowledge to produce and deploy nuclear and biological weapons that may not be optimal, but it can still cause horrendous destruction. It is mostly held in check not by insufficient S&T knowledge but by the lack of enriched uranium or a means for delivering biological agent over a wide area. Nevertheless, as demonstrated by the effects of modest-scale agent delivery by mail, a few letters containing anthrax sent by one (probable) individual can cause multiple deaths, panic, and economic calamity (FBI, 2010). Similar disruptive attacks on food supplies, electrical grids, key modes of transportation, and economic mechanisms such as stock markets are possible, if not likely, targets that S&T may enable.

MST 655, Advanced Conventional and Non-Conventional Weapons. This course should include a discussion of the importance of software in both advanced conventional and non-conventional weapons. Software subsystems are an important underlying technology of weapons systems. Furthermore, the precision of weapons is increasingly a function of software, hardware, and firmware elements. It is increasingly important for weapons analysts to understand the underlying logic that guides and controls weapons, both from the platform and within the warhead. Additionally, new weapons are increasingly developed and tested virtually, before any physical evidence is available for either collection or observation. As such, software simulation environments become an important element of new weapons discovery.

Finding 18: The MST 655 course related to the importance of software in weapon systems has a shortcoming.

Recommendation 18: The National Defense Intelligence College should include software design and simulation software in MST 655, with an emphasis on the use of these tools for the design and testing of new systems.

Another area worth covering in the course would be autonomous and semi- autonomous weapons systems such as UAVs. These systems are changing the way we fight on the battlefield and how we think about tactical warfare. While relatively new, these systems are being rapidly developed by many countries.

Finding 19: The syllabus for MST 665, The Biological Threat, is a thin outline compared with the substance in other sections in the concentration area such as MST 667, The Nuclear Threat.

Recommendation 19: The National Defense Intelligence College should greatly expand the content of MST 665 and label it as "under development" until it is improved.

Emerging and Disruptive Technologies

The NRC report *Persistent Forecasting of Disruptive Technologies* discussed in detail current forecasting methodologies, the nature of disruptive technologies, and the characteristics of disruptive technologies (NRC, 2010). Many researchers now see the importance of tracking technologies not only for how they are originally used, but how they are used by different groups of people. That NRC report stated as follows:

> The value of technology forecasting lies not in its ability to accurately predict the future but rather in its potential to minimize surprises. It does this by various means:
>
> - Defining and looking for key enablers and inhibitors of new disruptive technologies,
> - Assessing the impact of potential disruption,
> - Postulating potential alternative futures, and
> - Supporting decision making by increasing the lead time for awareness. (NRC, 2010)

S&TI analysts will need to understand how various technologies fit together and how out of the box thinking may affect or change whole industries very quickly. This is not a new thought. As Arthur Schopenhauer stated over 100 years ago:

> The task is not so much to see what no one yet has seen, but to think what nobody yet has thought about that which everybody sees.

The cultural lenses that overlay expectations color the ability to "see" potential applications and uses. It is more important to study an approach to correctly defining these potential uses than it is to review as many technology areas as possible.

Finding 20: The Emerging and Disruptive Technologies concentration area includes a very ambitious list of topics. The scope of problems is huge and rapidly changing. Employing competent instructors for these evolving topics will be challenging.

Recommendation 20: The National Defense Intelligence College should reduce the number of topics in the Emerging and Disruptive Technologies concentration to a manageable set that provides core competencies across a broad range of topics without being overly ambitious in terms of content.

Finding 21: The Emerging and Disruptive Technologies concentration area lacks a session on the innovative use of technologies. Obvious examples of this are flying airplanes into buildings and using the mail to distribute bioweapons-grade materials.

Recommendation 21: The National Defense Intelligence College should add a session to an existing course or create a new course on the innovative use of technologies to create threat capability.

Information Operations and Cyber

Cyber is a difficult area to cover: technology changes very quickly and new innovations are adopted globally at a dizzying pace. Information Operations and Cyber Intelligence is a very complicated topic area that spans not only raw IO & Cyber but also Cyber as a means to conduct other forms of ST&I. In fact, the basic vocabulary of IO does not appear to be addressed in the concentration which would allow leaders to understand at least the categories, significance, and components of modern cyber warfare and information operations. Furthermore, the areas of information operations and of cyber intelligence are in fact very different areas, with different skill sets, different objectives, and different candidate students.

Finding 22: The Information Operations and Cyber Concentration requires a precise definition of prerequisite knowledge for the students and for the instructors in order to create a slate of courses that are meaningful. Understanding what the graduate should be able to do is critical to defining both the learning objectives and the prerequisite knowledge requirements.

Recommendation 22-1: The National Defense Intelligence College should emphasize the development of a student's ability to learn, adapt, and deal with revolutions in cyberspace rather than emphasizing the learning of specific approaches or techniques that might work today but be obsolete within a year.

Recommendation 22-2: National Defense Intelligence College students focusing in the Information Operations and Cyber concentration should be required to have an appropriate undergraduate level of education, or life skills associated with information operations and cyber, or training appropriate to the learning objectives.

MST 680, Information Power and National Security. As the Internet increasingly becomes a conduit for both power projection and competition, all of the students must have prior knowledge so that this course can be conducted at the graduate rather than the introductory level. Students need an overview of how the Internet works and of how typical computer and IP networks are designed and must know about the growing dependency of everyday devices, appliances, systems, and infrastructure on IP.

Finding 23: The class should pay special attention to cloud-based architectures as well as mobile devices. Students should understand the basic strategies used to attack networks and have an understanding of defensive measures. There should be discussion of how the Internet can multiply the power of individual ideas and how this can be seen as threatening to other nation states.

While week 6 is devoted to Legal considerations in Information Operations, it is extremely important to discuss the various intelligence general counsel interpretations of existing laws, policies, and restrictions. In many cases, the agencies do not have the same interpretations, especially on issues such as U.S. persons, identity, data handling, covert action, and the active use of social media. Additionally, it is important to cover the role of standards, agreements, treaties in cyberspace and the strategies other nation states such as China, Russia, and the European Union are trying to reduce the United States influence and dominance in cyberspace. There are various materials available from the U.S. government, such as through the Joint Staff publications, that can be used to illustrate these issues.

Recommendation 23: The National Defense Intelligence College should add to MST 680 materials such as the Joint Information Operations Planning handbook and doctrine manuals for electronic warfare.

MST 681, Propaganda, Persuasion, and Influence. New social media technologies such as Facebook, Twitter, and Google are having a global impact. Recent examples of events that show the power and influence of social media include the 2009 Iranian elections[1], the Arab Spring of 2011[2], and the Great Firewall of China[3]. Analysts need to understand how social media is being used, particularly in coordinating and spreading political and social agendas. Further, they need to understand how social media are used to position and reinforce ideas and to generate and mobilize large portions of the populace around those ideas, particularly youth. A significant challenge lies in how social media are used in intelligence: At times social media blur the line between analysis and collection, and the potential for crossing that line needs to be deeply appreciated by members of the intelligence community. For example, active participation in social media could trigger specific rules and regulations that restrict how an analyst may operate, even with the most benign intentions, because some actions might be considered to fall into the realm of covert action. There is a distinct difference between reading a blog and friending or tweeting: One set of actions is passive while the other is active. There are challenges in how analysts and leaders manage and constrain their interactions with social media while still taking advantage of access to useful information.

Additionally, S&TI analysts and managers need to understand meta-social media issues, including the mechanisms of how usage is measured, how activities are tracked, the measurement of impact, assessing how social media is actually being used, and the identification of supernodes.

The emerging area of social media may prove to be too large to fit into one course; it might indeed need to be distributed across several courses. Not only are the preceding aspects of great importance, but there are also subtleties that need to be covered in the curriculum—including how different cultures use social media differently, how social media can be overwhelming traditional media at the same time they are feeding it, and how social media postings go "viral" (including characteristics of viral postings).

---

[1]See http://mashable.com/2009/06/21/iran-election-timeline/

[2]See http://blogs.reuters.com/great-debate/2011/02/16/digital-media-and-the-arab-spring/

[3]See http://www.datelineshanghai.com/scaling-the-great-internet-wall/

Finding 24: The area of social media appears to be neglected both in MST 681 and in the curriculum in general.

Recommendation 24-1: The MST 681 course should cover the use of social media and the importance of the social graph to show how they are used to facilitate nation-state protest, revolution, and social consciousness. The curriculum should cover the power of crowd sourcing, the effects of the digital exhaust, and social network tools such as Facebook, Google, YouTube, and Twitter. The class should also discuss Internet use to influence and shape culture and society.

Recommendation 24-2: The National Defense Intelligence College faculty should consider segmenting the challenges and learning objectives associated with covering social media and parceling out the learning needs across several different courses.

Finding 25: The backgrounds of the instructors for the Information Operations topics are critical to ensure that the material taught is deeper than a technical seminar from a vendor.

Recommendation 25: The National Defense Intelligence College should incorporate relevant open-source information on cyber-attacks for educational purposes and for stoking the imagination of students. Candidate material should include numerous technical analyses readily available online that cover recent cyber-incidents such as the Farewell Dossier, the Stuxnet worm, the Phishing Scheme against RSA, and the anonymous attack on HBGary, as well as future incidents as they occur.

Finding 26: The Information Operations syllabi reads more like political science focused syllabi than science or technology focused syllabi. For example, learning objectives are very high level and include phrases such as "gain an appreciation for" and "gain an understanding of." The basic vocabulary of this concentration area does not appear to be addressed so that it would allow leaders to understand at least the categories, significance, and components of modern cyber warfare and information operations.

Recommendation 26: The National Defense Intelligence College should repurpose the course content to be much more focused on the underlying science and technology and introduce people to the concept of cyber by exposing them to widely available hacking training for hands-on knowledge of systems compromise and manipulation.

Denial and Deception

The Denial and Deception concentration area is a well-developed program of study that has existed for quite some time. The courses for the MS&TI program have been repurposed from the existing curriculum for MSSI. Unfortunately, the resulting program of study that appears to be light on science and technology while strong in policy analysis.

Finding 27: There is little discussion of the implications of communications technology. The use of black, gray, and white propaganda and psychological operations is conspicuously absent. (This may be because it is in sensitive elements of the program.) Social media, which can be easily used for denial and deception, are not included in the provided material.

Recommendation 27: The National Defense Intelligence College should add the following content to the syllabi for MST 664:

Communications are now nearly instantaneous, ubiquitous, costless, anonymous, secure, and unlimited band-width. This enables worldwide and rapid flows of information, images, and money for good and ill. "In the global community, this leveraging power of information technology provides previously ineffective marginalized groups with additional means for both direct challenges to authority as well as indirect media for intra-group communications, planning, and management. Recent cases include the B-92 radio station in Bosnia[4], the use of the Internet by the Chiapas revolutionaries to marshal global attention to their cause[5], the use of tape-recorded messages by the Ayatollah Khomeini while in exile in Paris, the use of Twitter in the Iranian election protests, and the use of Facebook during the Egyptian protests[6]. These examples illustrate the breakdown of boundaries, both physical and customary" (Ryan, 2011).

Finding 28: The current Foreign Denial and Deception concentration does not distinguish between the MSSI and MS&TI programs. Only one course, MST 604, is different, and MST 613 would be optional in MS&TI. Additionally, there is a dearth of technology and science in the syllabi.

Recommendation 28: The National Defense Intelligence College should incorporate a stronger background in science and technology in the Denial and Deception concentration area by requiring one or more additional courses that focus on S&T-based based deception.

CONCLUSION

The need for the NDIC's new degree program was found to be justified. The user population reflected strong support for this effort. An examination of the program also revealed that it meets the four criteria referenced: it is necessary, it is unique, it meets general academic practice, and the administrative processes support academic freedom.

REFERENCES

Published

Choffnes, Eileen R., Stanley M. Lemon, and David A Relman. 2006. "A brave new world in the life sciences: The breadth of biological threats is much broader than commonly thought and will continue to expand." *Bulletin of the Atomic Scientists* September/October: 26-33.

Crawley, Edward, Johan Malmgvist, Soren Ostlund, and Doris Brodeur. 2007. *Rethinking Engineering Education: The CDIO Approach.* Springer Science+Business Media, pp. 40 and 113.

Engineering Trends. 2006. "Historical trends and near-term predictions of statistics on degrees, enrollments and research expenditures for engineering education in the United States." *Engineering Trends Quarterly Newsletter.* Summer 2006.

---

[4]See for example "Preserving the Free Flow of Information on the Internet: Serbs Thwart Milosevic Censorship" at http://www.usip.org/events/preserving-free-flow-information-internet-serbs-thwart-milosevic-censorship-round-two.

[5]See for example "Rebellion in Chiapas: insurrection by Internet and public relations" by Jerry W. Knudson from Temple University, available online at http://www.bsos.umd.edu/aasp/chateauvert/rebellio.pdf.

[6]See for example "The dark side of Internet for Egyptian and Tunisian protesters", which discusses many of these events. By Evgeny Morozov for the Globe and Mail, published Jan. 28, 2011. Available online at http://www.theglobeandmail.com/news/world/africa-mideast/the-dark-side-of-internet-for-egyptian-and-tunisian-protesters/article1887170/print/.

Federal Bureau of Investigation (FBI). 2010. *Amerithrax or Anthrax Investigation Summary.* February 19. Available online at http://www.fbi.gov/about-us/history/famous-cases/anthrax-amerithrax/amerithrax-investigation.

Friedman, Thomas L. 2010. "Can't keep a bad idea down." *The New York Times*, October 26. Available at http://www.nytimes.com/2010/10/27/opinion/27friedman.html. Last accessed on July 20, 2011.

Intelligence Science Board (ISB). 2006. *Intelligence Science Board Task Force Report on the Intelligence Community and Science and Technology: The Challenge of the New S&T Landscape.* November. Washington D.C.: Office of the Director of National Intelligence. Available at http://www.fas.org/irp/dni/isb/landscape.pdf. Last accessed on September 1, 2011.

National Academy of Sciences, National Academy of Engineering, Institute of Medicine (NAS, NAE, IOM). 2010. *Rising Above the Gathering Storm, Revisited: Rapidly Approaching Category 5.* Washington, D.C.: The National Academies Press. Available at http://www.nap.edu/catalog.php?record_id=12999. Last accessed on July 12, 2011.

National Defense Intelligence College (NDIC). 2011a. *NDIC University Catalog 2011-2012.* Washington, D.C.: Center for Strategic Intelligence Research.

NDIC. 2010. *National Defense Intelligence-2010 Annual Report.* Available at http://www.dia.mil/college/. Last accessed on July 12, 2011.

National Research Council (NRC). 2011a. *Intelligence Analysis: Behavioral and Social Scientific Foundations.* Washington, D.C.: The National Academies Press. Available at http://www.nap.edu/catalog.php?record_id=13062. Last accessed on July 15, 2011.

NRC. 2011b. *Intelligence Analysis for Tomorrow: Advances from the Behavioral and Social Sciences.* Washington, D.C.: The National Academies Press. Available at http://www.nap.edu/catalog.php?record_id=13040. Last accessed on July 15, 2011.

NRC. 2010. *Persistent Forecasting of Disruptive Technologies.* Washington, D.C.: The National Academies Press.

NRC. 2006. *Globalization, Biosecurity, and the Future of the Life Sciences.* Washington, D.C.: The National Academies Press. Available at http://www.nap.edu/catalog.php?record_id=11567. Last accessed on July 15, 2011.

NRC. 2005. *An International Perspective on Advancing Technologies and Strategies for Managing Dual-Use Risks: Report of a Workshop.* Washington, D.C.: The National Academies Press. Available at http://www.nap.edu/catalog.php?record_id=11301. Last accessed on July 15, 2011.

Office of the Director of National Intelligence (ODNI). 2008. *Intelligence Community Directive Number 652, Occupational Structure for the Intelligence Community Civilian Workforce.* ICD 652, April 28. Available at http://www.fas.org/irp/dni/icd/icd-610.pdf. Last accessed on September 1, 2011.

Ryan, Julie (ed.). 2011. *Leading Issues in Information Warfare and Security Research,* Volume 1. Reading, England: Academic Publishing International. Pages xiv and xv.

The White House, Office of the Press Secretary. 2009. "President Obama launches "Educate to Innovate" campaign for excellence in science, technology, engineering & math (STEM) education. Press Release, November 23. Available at http://www.whitehouse.gov/the-press-office/president-obama-launches-educate-innovate-campaign-excellence-science-technology-en. Last accessed on July 12, 2011.

Unpublished

NDIC. 2011b. *National Academy of Sciences–Curriculum Review of the Master of Science and Technology Intelligence, May 2011*. Washington, D.C.: Center for Strategic Intelligence Research, NDIC.

O'Neil, Joseph. "The faculty's role in curriculum development." Presentation to the committee on May 13, 2011.

Studds, Susan, Brian Smith, Joe O'Neil, Gerry Sherrill, Duncan McGill. "National Defense Intelligence College – School of S&T Intelligence." Presentation to the committee on May 13, 2011.

Appendix

Appendix A

Biographical Sketches of Committee Members

Diane Griffin *(NAS, IOM), Chair,* is professor and chair of molecular microbiology and immunology and founding director of the Johns Hopkins Malaria Institute at the Johns Hopkins Bloomberg School of Public Health. She earned a biology degree from Augustana College in 1962, followed by M.D. (1968) and Ph.D. (1970) degrees from Stanford University. She interned at Stanford University Hospital between 1968 and 1970, before beginning her career at Johns Hopkins as a postdoctoral fellow in virology and infectious disease in 1970. After completing her postdoctoral work, she was named an assistant professor of medicine and neurology. Since then, she has held the positions of associate professor, professor, and, now, professor and chair. She served as an investigator at Howard Hughes Medical Institute from 1973 to 1979.

Dr. Griffin's research interest includes alphaviruses and acute encephalitis. She is also working on the effect of measles virus infection on immune responses in monkeys and in humans at the University Teaching Hospital in Lusaka, Zambia. In Zambia, she and her colleagues have been examining the effect of HIV infection on measles and measles virus immunization.

Dr. Griffin has been the principal investigator for a variety of grants from the National Institutes of Health, the Bill & Melinda Gates Foundation, and the Dana Foundation. She is the author or coauthor of a number of scholarly papers and articles, is the past president of the American Society for Virology, the Association of Medical School Microbiology Chairs, and the American Society for Microbiology. She is a member of the National Academy of Science, the Institute of Medicine, and the American Academy of Microbiology.

Julie J.C.H. Ryan, *Vice Chair,* is associate professor and chair of the Department of Engineering Management and Systems Engineering at The George Washington University. She holds a B.S. degree in humanities from the U.S. Air Force Academy, an M.L.S. in technology from Eastern Michigan University, and a D.Sc. in engineering management from the George Washington University. Dr. Ryan began her career as an intelligence officer, serving the U.S. Air Force and the Defense Intelligence Agency and then working a series of increasingly responsible positions throughout her career. Her areas of interest are information security and information warfare research. She currently serves as a member of the National Research Council's (NRC) Standing Committee for Technology, Insight—Gauge, Evaluate, and Review and was a member of NRC's Naval Studies Board from 1995 to 1998. She has had a distinguished career, having conducted several research projects and authored articles, book chapters, monographs, and a book in her focus area.

Brian Ballard is the director of product development for Berico Tailored Systems. Before this, he was the chief technology officer of MAV6, where he was involved in the development of emerging networking and embedded systems technologies for intelligence, surveillance, and reconnaissance (ISR) systems and applications in government and the military. He is a highly experienced professional in the field of national intelligence systems and computer engineering. Employed for more than 10 years by the National Security Agency, Mr. Ballard has dealt with all forms of data collection, dissemination, processing, and visualization. As a field operations officer at the NSA, he was a member and team leader in the Office of Target Reconnaissance and Surveillance. He also worked for 5 years as a global network vulnerabilities analyst. Mr. Ballard holds an M.S. in electrical and computer engineering and a B.S. in electrical and computer engineering from Carnegie Mellon University. He is currently studying for an M.S. in technology management and an M.B.A. at the University of Maryland in College Park.

Wesley Harris (*NAE)* is the Charles Stark Draper Professor and head of the Department of Aeronautics and Astronautics at Massachusetts Institute of Technology. His research focuses on theoretical and experimental unsteady aerodynamics and aeroacoustics; computational fluid dynamics, and the impact of government policy on procurement of high technology systems. Before this he served as the associate administrator for aeronautics at NASA. He has also served as the vice president and chief administrative officer of the University of Tennessee Space Institute. Dr. Harris earned a B.S. in aerospace engineering from the University of Virginia and M.S. and Ph.D. degrees in aerospace and mechanical sciences from Princeton University.

Kenneth A. Kress is a senior scientist for KBK Consulting, Inc., and a consultant for Booz Allen Hamilton, where he specializes in quantum information science and other technical evaluations and strategic planning for intelligence and defense applications. Some of his past clients include DARPA's Microsytems Technology Office, Noblis, Georgia Tech Research Institute, Mitretek Systems, Inc., and Lockheed Martin Special Programs Division. From 1971 to 1999 he worked in a series of positions at the Central Intelligence Agency's Directorate of Operations, Office of Development and Engineering, and, finally, its Office of Research and Development (ORD), first as a research and development manager, later as a program manager, and finally as an ORD senior scientist responsible for management support, the development of technical and strategic plans, and DOD interagency coordination for advanced technology. He is the inventor of the solid state neutron detector, for which he won an award in 1981. He holds a Ph.D. in physics from Montana State University.

Gilman Louie is a partner of Alsop Louie Partners, a venture capital fund focusing on the development of technology entrepreneurs. Earlier, he was president and CEO of In-Q-Tel, the venture capital group helping to deliver new technologies to the CIA and the intelligence community. Before helping found In-Q-Tel, Mr. Louie served as Hasbro Interactive's chief creative officer and as general manager of the Games.com group, where he was responsible for creating and implementing the business plan for Hasbro's Internet games site. Before joining Hasbro, he served as chief executive of the Nexa Corporation; Sphere, Inc.; and Spectrum HoloByte, Inc. As a pioneer in the interactive entertainment industry, Mr. Louie's successes have included the Falcon Fighting F-16 flight simulator and Tetris, which he brought over from the Soviet Union. Mr. Louie has served on the board of directors of Wizards of the Coast, Total Entertainment Network, Direct Language, and FASA Interactive. He is an active member of the Markle Foundation Task Force on National Security and the Information Age and is a member of the board of New Schools.org. Mr. Louie was chosen for his expertise in intelligence, threat analysis, and venture capital new technology start-ups.